Secret Universe
PATTERNS OF MOTION

by Dwayne Osterbauer

based on the discoveries of

Louis A. Osterbauer, M.S.
and
E.J. Muybridge

Secret Universe:
Patterns of Motion

copyright 2008 Triad Productions

First Edition

COMPANION DVD AVAILABLE SOON ON AMAZON.COM

Secret Universe

CONTENTS

What It's About ---------------------------------- 1
The Birth of Kinegraphy/
   Venus & Earth ---------------------------------- 3
Uranis & Neptune ---------------------------------9
Jupiter & Venus ---------------------------------- 11
Jupiter & Saturn ---------------------------------- 13
Planetary Kinegraphy Discovered --------- 15
Planet "X"? -- 17
Jupiter & Uranis ---------------------------------- 19
Earth & Mars ---------------------------------- 21
Mars & Jupiter ---------------------------------- 23
Relativity -- 25
Dante's Vision ---------------------------------- 29
Final Thoughts ---------------------------------- 33

What It's About

"...there is Music in the spacing of the Spheres."
(words of the ancient mathematician and philosopher Pythagoras)

Pythagoras could have meant those cosmic spheres we call planets. This book will reveal, in visual form, the "music" in the spaces between the planets.

This is not something mystical or metaphysical. There are provable, factual realities beyond visible matter and detectable energy -- beyond the things revealed by microscopes, telescopes, and radio arrays.

The reality we'll reveal in this book can only be exposed through a special process called kinegraphy, which we'll describe in the next chapter.

An amazing yet simple technique with the power to reveal at last Pythagoras' "Music of the Spheres".

The Birth of Kinegraphy/
Venus & Earth

To understand how these invisible things are revealed,
we must travel back to the 1870's.

photography pioneer E.J. Muybridge

In this era, there were no motion pictures, only still photography.
E. J. Muybridge was about to change all that. A friend was a horse racing enthusiast,
who made a bet that while galloping, there are moments when all four
of a horse's hooves are off the ground at the same time.

"Balderdash! Horses can't fly!", the others grumbled.

The only way to prove his point was to photograph a horse using a series
of still cameras set up along the track, to capture each moment.

When the opposition wasn't convinced by the stills, Muybridge invented a device that could show the photos in rapid sequence, plainly revealing the horse's hooves *in action*.

He called his contraption the "zoopraxiscope" (thought to be Thomas Edison's inspiration for inventing the motion picture projector). Public exhibitions of this amazing new technology became wildly popular.

Others soon expanded upon Muybridge's original techniques. Instead of projecting the stills in sequence, they experimented with overprinting the entire sequence onto a single photographic print or plate. It's not known when or by whom, but someone dubbed this process, "kinegraphy". Below is one of the more famous examples, titled, "standing broad jump".

But they didn't stop there. To help them analyze the motions, someone came up with the idea to lay a pane of glass over the photo, on which they'd draw lines connecting the joints of the body.

When we look at just the lines, we see the athlete's
pattern of motion, each line representing one moment in time.

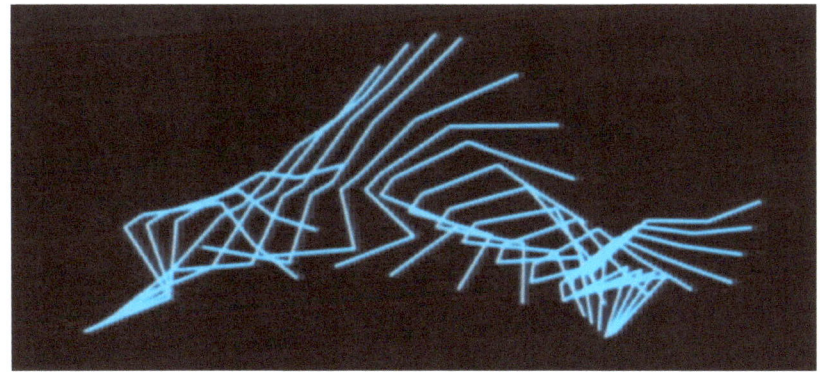

The trail of lines shows where each skeletal joint was
in relation to another at one moment in time, then another, then another...

A similar type of kinegraphy applies to *celestial* bodies. Here's Earth and Venus.

(not to scale)

Just as with joints of the body, we draw a straight line between these two planets.

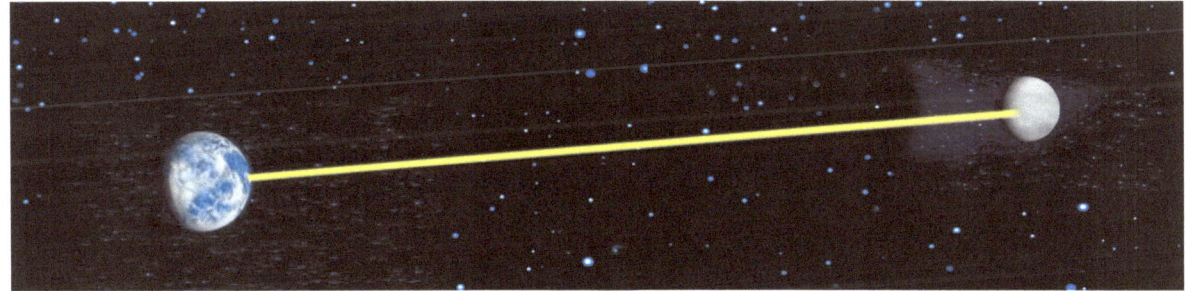

It's the actual line of sight when you look at Venus from Earth.

To reveal the pattern this generates, you need one more ingredient:
The MOTION of the planets as they journey around the sun.

On every-other Earth day, we draw one line between Venus and Earth.

As the planets travel, we leave behind a trail, showing where the lines of sight were on each of those alternate days.

(We're doing every OTHER day just to keep the design from looking too crowded.)

Here's the way it looks after six months' worth of lines...

...and a little over one year.

After five orbits (years) for Earth (eight orbits for Venus) the pattern is complete. The whole thing then repeats itself, each time shifting slightly clockwise.

Why the shift? Don't forget leap year! Earth's year is not precisely 365 days -- it's closer to 365 and one-quarter days, so by the time the design completes itself, it's all shifted a few degrees. The Venutian year is also not perfect.

Uranis & Neptune

Pluto used to be considered a planet, but most astronomers now agree it's something else, like a comet, a stray moon or asteroid. Anyway, it doesn't make for good kinegraphy patterns, because the angle of its orbit is tilted far out of line with the other planets.

So, let's travel inward, toward the sun. The next stop is Neptune, followed by Uranis, which has thin, "Saturn-like" rings, and is strangely tilted over on its side.

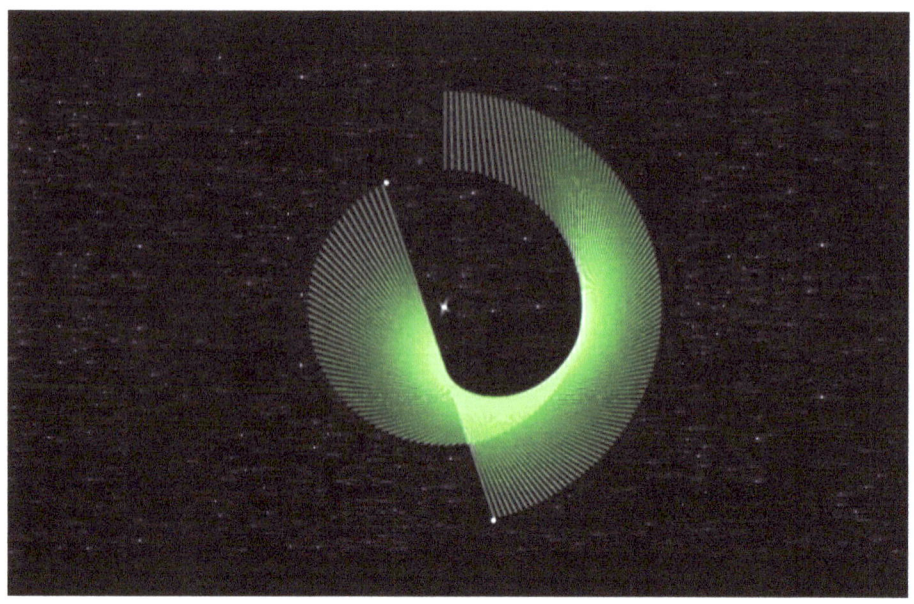

Their pattern starts out looking a lot like Venus-Earth's, but it forms only one heart shape instead of three. By the way, since these patterns are invisible, they have no actual color, so we can render them in any shade we like.

Jupiter & Venus

Jupiter is by far the largest of the planets. You could fit all the other planets inside of Jupiter, and have room enough left over for a whole second set.

It's literally billions of miles from Jupiter to Venus, making this one of the largest patterns.

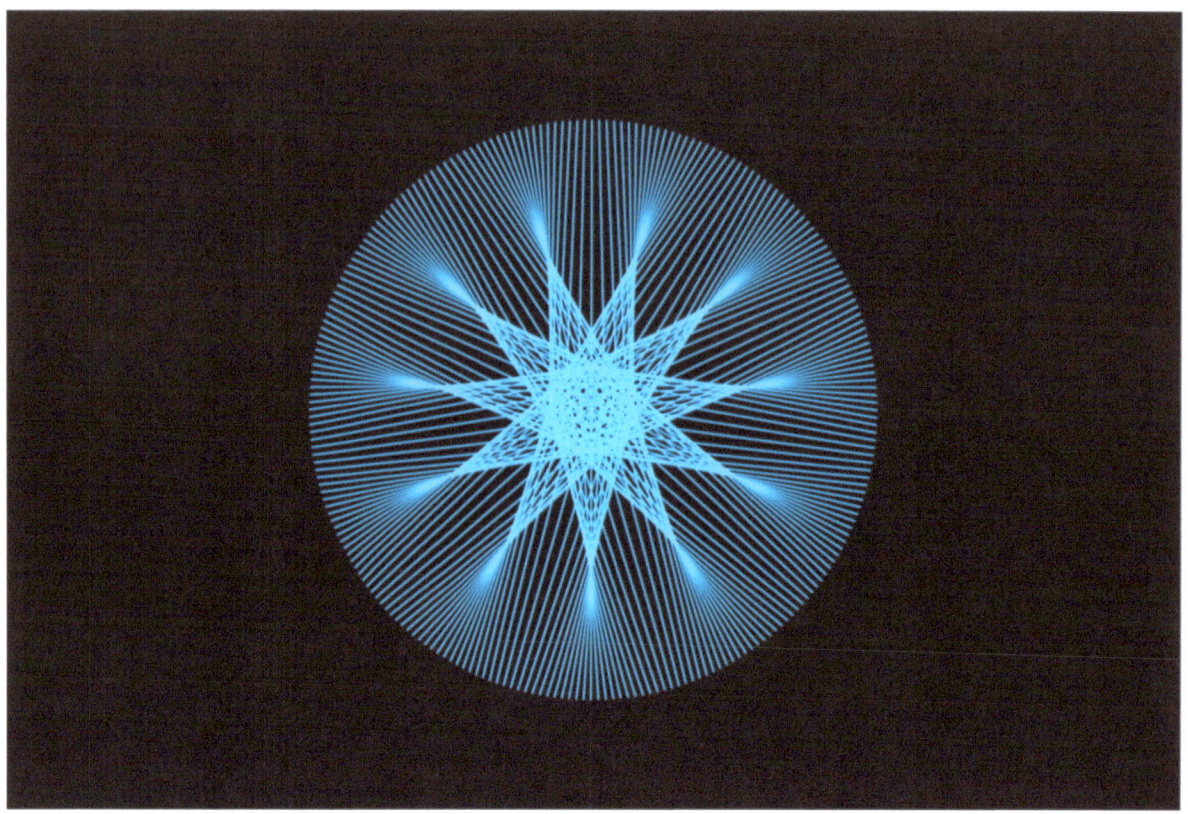

Jupiter-Venus

This is the second type of pattern: a star shape.

On any orbital kinegraphy, if the difference in the size of the planets' orbits is smaller than a certain proportion, you get inward dips or hearts (like Earth-Venus).

If it's a larger difference, you get outward points or stars, as in the example above.

Jupiter & Saturn

We've already met Jupiter, the largest planet.

Next largest is Saturn, whose famous rings look solid and semi-transparent. They're actually made up of trillions of separate chunks of debris, but their close spacing at this distance creates the familiar illusion. Jupiter and Saturn are close neighbors...

"Close", meaning over 400 million miles (700 million km) when they're on the same side of the sun together, billions of miles apart when they're on opposite sides.

Strange fact: this pattern is almost identical to the kinegraphy between the two innermost planets, Mercury and Venus.

Planetary Kinegraphy Discovered

It was almost ninety years after Mr. Muybridge's innovations prompted the discovery of earthbound kinegraphy.

Louis A. Osterbauer with his grandchildren in the 1980's

Mathematician Louis A. Osterbauer was looking for something to spice up a geometry lecture. An amateur astronomer, he decided to plot the retrograde loops of the planets. What are "retrograde loops"?

If you plot the position where a planet appears every night, from our position it sometimes looks like it's looping around backward, before continuing across the sky.

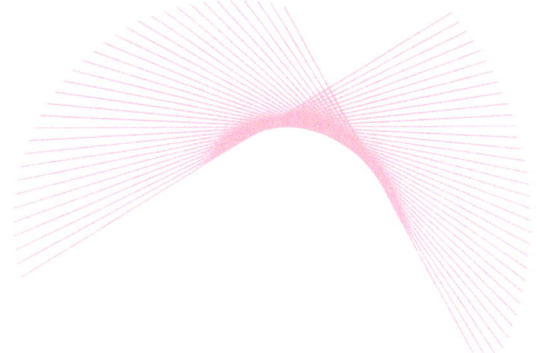

To see exactly how this works, he took the example of Earth and Venus and plotted their positions for every-other night over one Earth year (see page 6 of this book).

Imagine his surprise when he saw the pattern this produced. Only, it wasn't the design you see on page 7. Remember, he plotted only ONE Earth year, so instead of the three inner loops, he got...

...THIS!

That's right, a one year kinegraph of Earth's view of Venus, the ancient goddess of love, actually makes a heart!

Was he the only person ever to discover this phenomenon? Possibly not, but in all the years since, the author has never found a single other example.

Planet "X"?

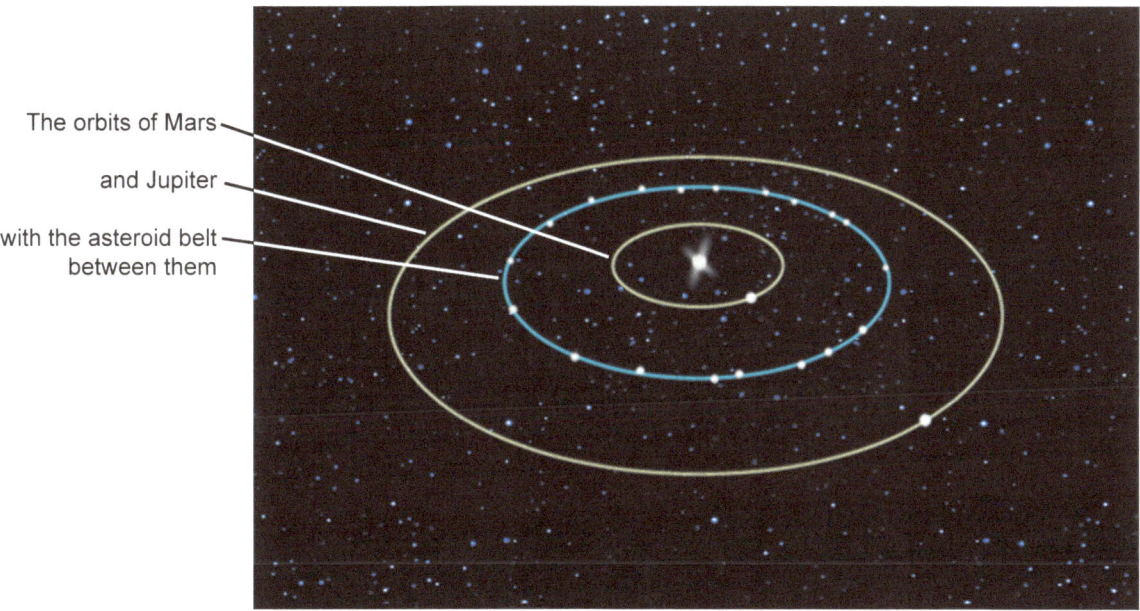

The orbits of Mars and Jupiter with the asteroid belt between them

There's a pattern to the spacing of the planets. A progression of orbit sizes. But there's a mysterious gap in this sequence. You'd expect another planet between the orbits of Mars and Jupiter, but instead all you'll find is a gigantic "belt" of asteroids. Fragments of rock, ranging in size from less than an inch, all the way up to the largest, called "Ceres".

A cold chunk of stone and ice, almost 600 miles across.

A photo of Ceres taken by the Hubble Space Telescope. Until recently, it was thought to be oblong instead of sperical as this picture demonstrates.

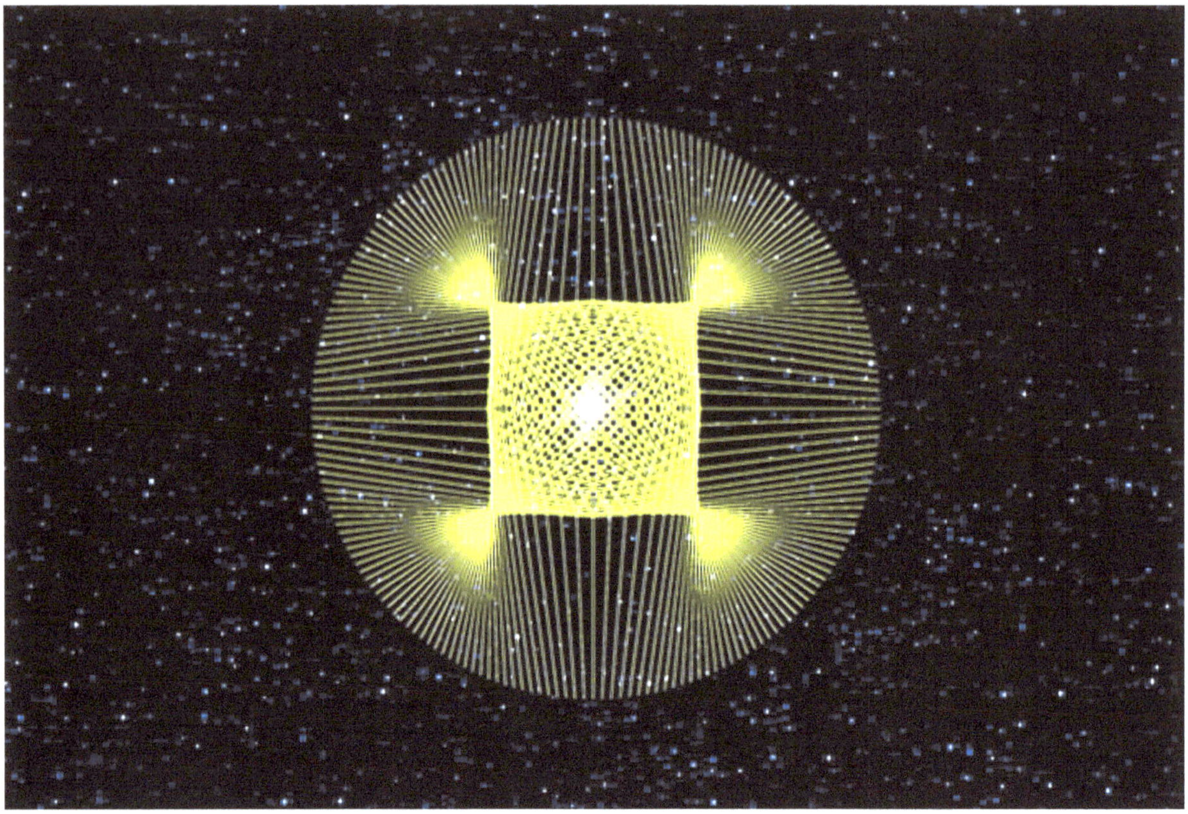

Here's the kinegraphic pattern it makes with Earth.

There are various theories why there's no actual planet in place of Ceres. Perhaps one existed long ago, which was destroyed. Or maybe the powerful forces of Jupiter's gravity kept one from forming, or possibly something else.

To Professor Osterbauer, it was just a case of a square peg trying to fit into a round hole.

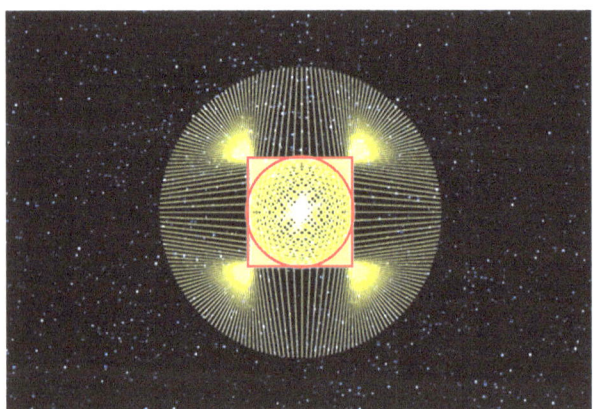

(mathematician's humor)

Jupiter & Uranis

Not only is Jupiter the largest planet, it also hosts the biggest planetary storms in the solar system, including hurricanes thousands of miles across --that last hundreds of years!

While Earth's axis is tilted over just a few degrees, giving us our seasons, Uranis is almost lying over on its side, spinning from north to south instead of west to east.

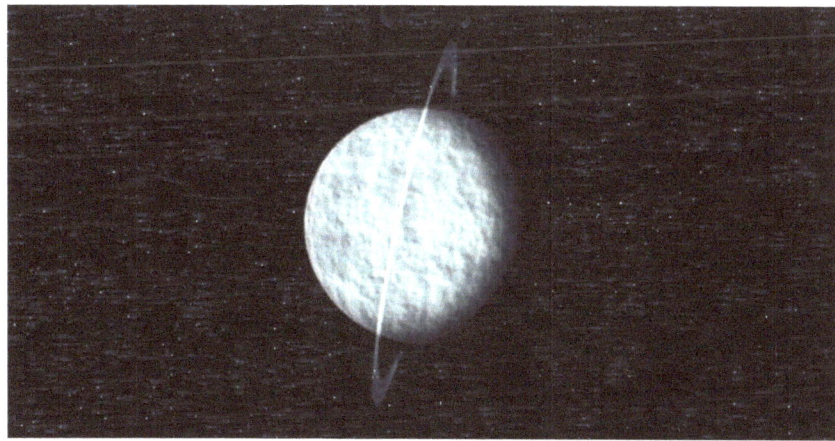

Religious symbol? Badge of authority?
Or just a coincidence.

Earth & Mars

This design starts out looking similar to Earth-Venus.

But after a couple of orbits, we begin to see that it's going to become more complex than that.

In fact, Mars-Earth has the most complex planetary kinegraphy of all.

Strange, perhaps, that this lovely, serene design is formed by Mars, named for the ancient god of war.

Mars & Jupiter

Again, coincidence? Mathematical probability? Intentional clue?
You decide.

Relativity

Einstein had a famous theory, which for our purposes we'll put in simple terms.

(Scenes from *Going Home* courtesy Daniel Four Motion Pictures)

If you're the driver of the car, the girl seems stationary, relative to your position. Your "point of reference".

If you're outside, standing by the road, that becomes your new point of reference, and the girl (and everything in the car) are moving pretty fast.

It's the same for the planets.

As a planet orbits, it forms a full circle, relative to the sun. But the sun's in motion too, relative to the whole galaxy. So, from that point of reference the planet doesn't make circles. It makes loops.

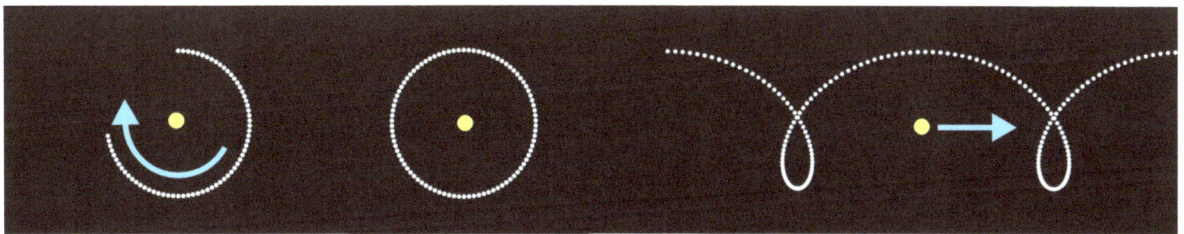

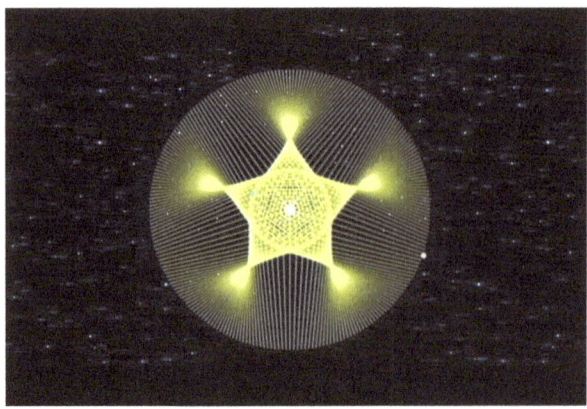

Above, Jupiter-Mars as seen from a stationary point of reference.
Below, the same pattern as it unfolds from a vantage point outside the solar system.

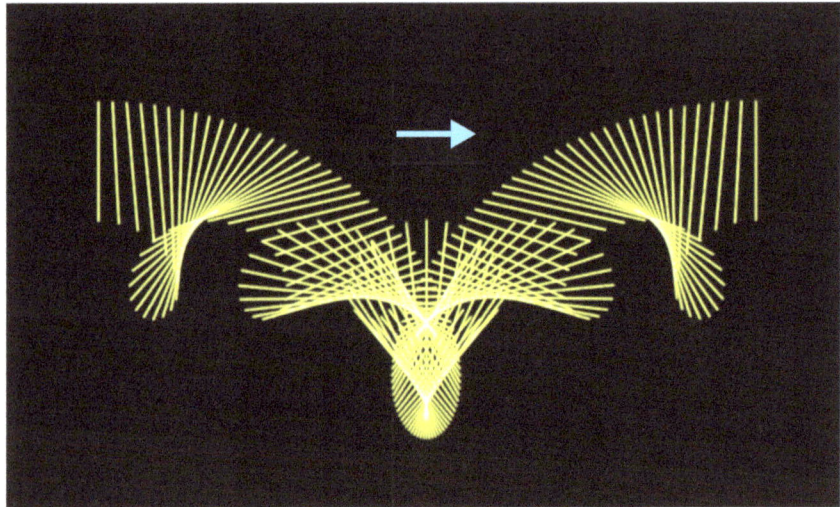

But relativity doesn't end there.
Our Milky Way galaxy is also in motion, on its axis:

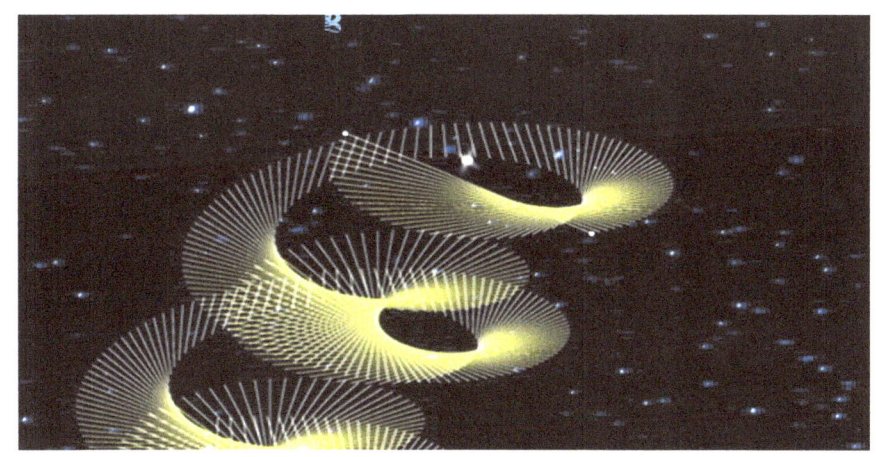

So, from outside the galaxy, you get patterns like the one above.

Here are some more examples of relativistic kinegraphy:

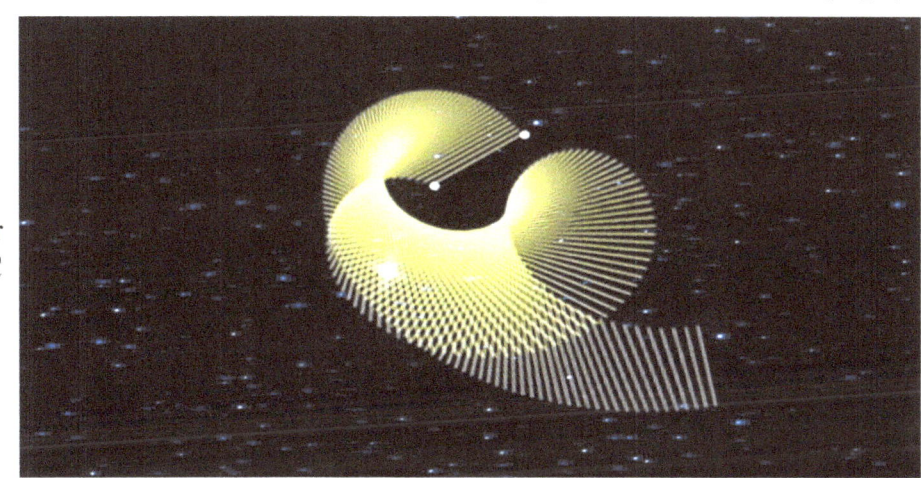

Saturn-Jupiter seen from axis C

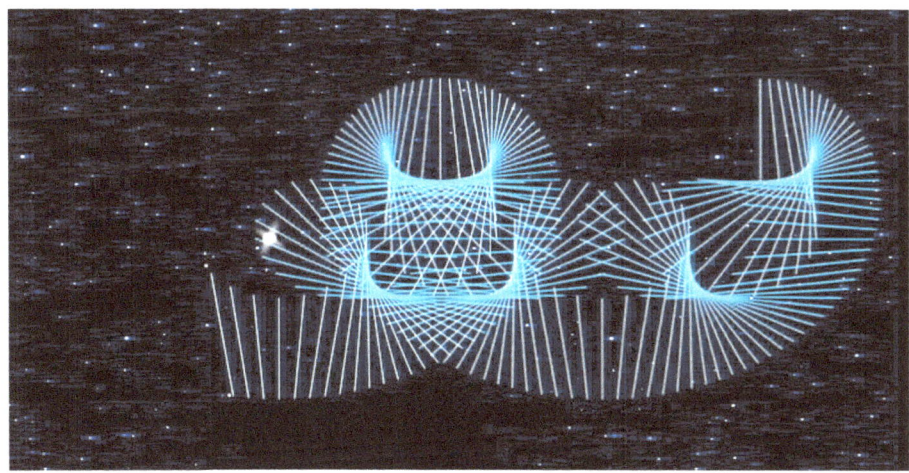

Ceres-Earth seen from axis B

Dante's Triad

In the early Renaissance, in Florence, Italy,

a young poet was contemplating the nature of heaven.

He had a vision, which years later he would immortalize in poetry.

He saw a celestial corridor,

leading to a bright light,

with angels flying toward it.

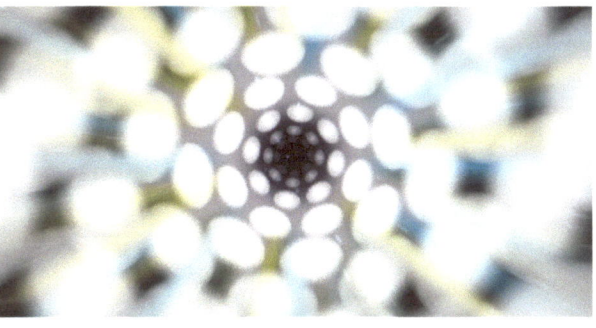

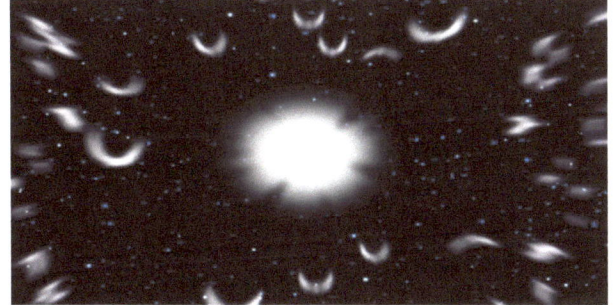

As he flew within the veil of light, he saw three circular patterns, each perfect on its own, and each in harmony with the other two.

A hallucination? A poetic fantasy?
Or does something resembling this actually exist?

So far, we've only shown the kinegraphy between two spheres at a time.
But what if you include a third?

Here are the three inner planets, starting with Mercury, closest to the Sun, then Venus, and Earth. Instead of just one line of sight, there are now three.

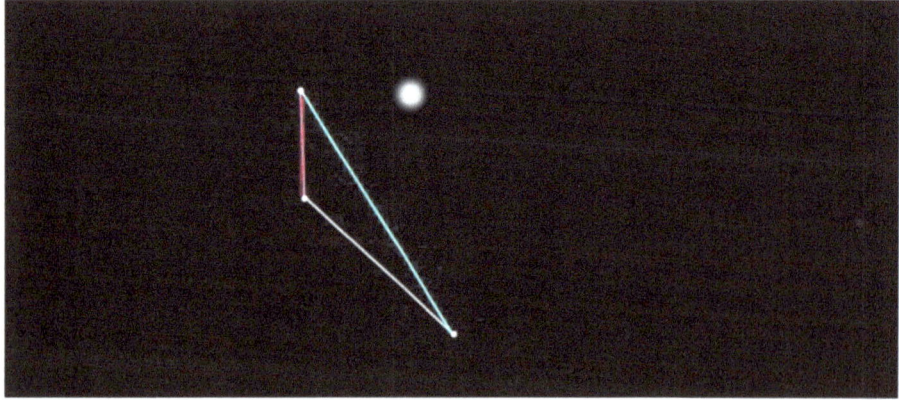

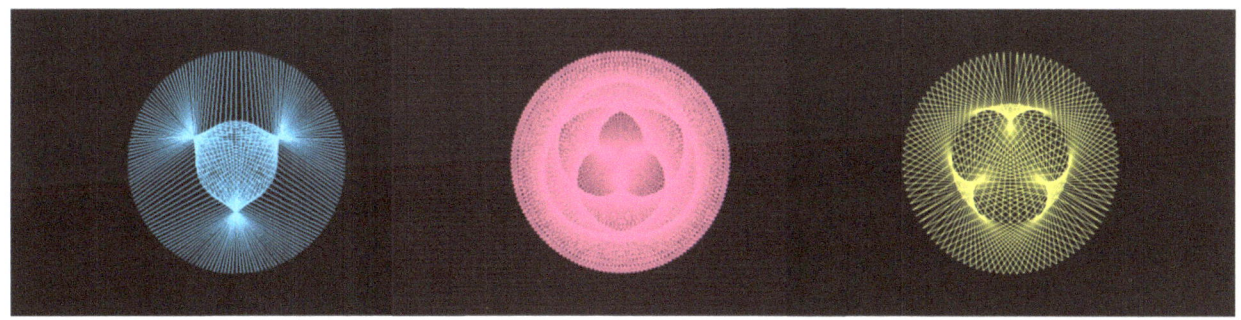

What you get is a perfectly harmonious triad composed of Earth-Mercury, Earth-Venus, and Mercury-Venus (in order, left to right above).

Is this Dante's vision? No one knows. But depending on your interpretation, the similarity is remarkable.

Final Thoughts

Everyone sees things differently.

When you look at these patterns, do you see evidence of a Higher Power? An intelligent Universe? Mathematical principles ...or coincidence?

We tend to interpret any new discovery in terms of our expectations. And our expectations are largely determined by our needs. What we want or need to believe.

A universal human need is to see order and harmony in the Universe, and that is perhaps why so many people find these patterns so appealing. It's a feeling that perfection does exist.

If you would prefer to leave this impression intact, please stop reading here. On the next page is a confession which may spoil that feeling (a little).

The Spoiler?

Although Louis Osterbauer was an athiest, he too needed to see perfection and symmetry in the universe. So, when he plotted the orbital patterns, he rounded off his figures to make everything come out even, which resulted in kinegraphy that looked perfect and symmetrical, as in the top illustration below.

But in reality, if you use the actual specifications of these two planets' orbits, it comes out looking more like the bottom picture.

"So what? It's still beautiful", you might say.
It all depends on your needs. If you're looking for perfection, sorry for the disappointment.

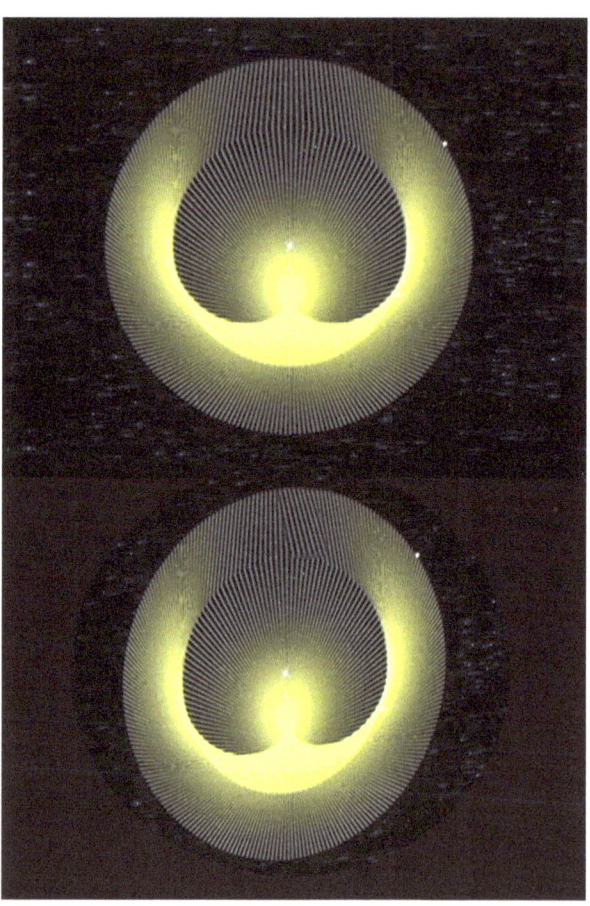

"Alas, perfection is not for this world."
--*Arturo Toscanini,
renowned symphony conductor*

www.ingramcontent.com/pod-product-compliance
Lightning Source LLC
Chambersburg PA
CBHW051111180526

45172CB00002B/859